All Kinds of CLOUDS

Written By: Anna DiGilio

All rights reserved. No part of this publication may be reproduced, distributed, or transmitted in any form or by any means, including photocopying, recording, or other electronic or mechanical methods, without the prior written permission of the publisher, except in the case of brief quotations embodied in critical reviews and certain other noncommercial uses permitted by copyright law.

For permission requests, write to the publisher:
Laprea Publishing
info@lapreapublishing.com

Website: www.GuidedReaders.com

ISBN: 978-1-64579-594-0

© 2019 Anna DiGilio

Photo Credits:
Cover, Title Page: Depositphotos; Despotoll. 3, 7 (bottom), 10: Depositphotos; Mihalec. 4: Depositphotos; Christianhinkle. 5: Depositphotos; Shmeljov. 6: Depositphotos; Drizzd. 7 (top): Depositphotos; MajaFOTO. 7 (center): Depositphotos; Searagen. 8: Depositphotos; SimpleFoto. 9: Shutterstock; Vvoe. 11: Shutterstock; PRESNIAKOV OLEKSANDR. 12: Depositphotos; Uroszunic.

TABLE OF CONTENTS

Kinds of Clouds ....Page 5

Watch the Clouds ....Page 12

Glossary..Page 13

Look up in the sky. What do you see? Clouds!

Kinds of Clouds

There are many kinds of clouds. There are over one hundred kinds!

There are ten major types of clouds. <u>Scientists</u> group them. Groups are based on <u>height</u>.

There are low clouds. There are middle clouds. There are high clouds.

Cumulus clouds float low. They are big. They are white. They are <u>puffy</u>.

Stratus clouds hang low. They are thin. They are flat. They are gray. They mean rain.

Cirrus clouds fly high. They are <u>wispy</u>. They are curly. They mean sunny days.

Cumulonimbus clouds are special. They are huge! They are so big. They are in low, middle, and high parts of the sky all at once! Watch out! They mean storms.

Watch the Clouds

Clouds are fun to watch. Look for shapes. Clouds are helpful. They tell you about the <u>weather</u>.

GLOSSARY

<u>heights</u>
high places or areas

<u>puffy</u>
soft, rounded, and light

<u>scientists</u>
people who have training in, study, and work in, a field of science

<u>weather</u>
the state of the air at a certain time and place, such as hot, cold, clear, cloudy, or rainy

<u>wispy</u>
fine; feathery